"COMPASSION IS NOT A CRIME"

The Trial of Anita Krajnc

Kyle Ferguson

CONTENTS

PROLOGUE

"When the suffering of another creature causes you to feel pain, do not submit to the initial desire to flee from the suffering one, but on the contrary, come closer, as close as you can to him who suffers, and try to help him."
—Leo Tolstoy, from A Calendar of Wisdom

It was a warm day outside, about 26 degrees Celsius. Inside the trailer it was much hotter.

There were small open windows on the sides, but they didn't help much. He was crammed in with too many others. It was dirty, and there was little light, and he was panting in the heat. His mouth was dry. He was thirsty. They all were. And they were being taken to be killed.

The truck pulling them slowed down, then stopped. A traffic light.

Some people ran up. "Can you give this guy some water?" one said.

A woman reached in through one of the small windows with

a bottle of water. He reached his head up to have some. He heard the driver running towards them, shouting: "Don't give him anything. Do not put water in there."

"If they are thirsty," the woman said, "Jesus said if they are thirsty, give them water."

"No," the driver shouted. "You know what, these are not humans, you dumb frickin' broad! Hello!" He started thumbing numbers on his phone. "You know what, now we're going to call the cops."

"Have some compassion," she said. "Have some compassion."

He kept dialling. She reached in with the water to give them some.

He pointed at the bottle. "What do you got in that water?"

"Water."

"No. No. How do I know?"

"You can test it. I'll give you a sample."

"Don't put it in there again."

She gestured towards the panting occupants of the trailer. "If this pig is thirsty, they'll have water."

"You do it again and I'll slap it out of your hands."

"Go ahead, if you want assault charges." She turned to her companion. "Film this. If he wants to have assault charges, go ahead." She reached in and squirted water on the pig's snout, and he drank a few sips eagerly.

That's where the video ends.[1] But we know what happened after that.

The truck, with its trailer—53 feet long, 8 feet wide, 3 decks high, containing 200 pigs raised on Van Boekel Hog Farms[2]—continued into Fearmans Pork slaughterhouse, where every one in the trailer was stunned, cut open, killed, cut to pieces, and packaged.

And the owner of the truck, Eric Van Boekel, made a complaint against the woman, Anita Krajnc, for giving them a last drink of water as they were taken in the heat to be killed. She was charged with criminal mischief and faced up to ten years in jail.[3]

This is the story of her trial.[4]

---

[1] Murphy, J. (2015, November 30). Canada woman faces 10 years in prison for giving pigs water on hot day. Retrieved February 17, 2020, from https://www.theguardian.com/world/2015/nov/30/canada-woman-10-years-prison-for-giving-pigs-water

[2] Data given by Eric Van Boekel during the trial.

[3] Preliminary trial date for woman charged after giving water to pigs | CBC News. (2015, November 4). Retrieved March 2, 2020, from https://www.cbc.ca/news/canada/hamilton/news/preliminary-trial-date-for-woman-charged-after-giving-water-to-pigs-1.3303599

[4] For the sake of presenting the details clearly and concisely, the trial details have been condensed. The witnesses presented are not the only persons who testified during the trial, not all of their testimony is included, and they did not all testify in the order in which their testimony is presented. The quoted testimony does not contain any added words (except as indicated in square brackets for clarification) or paraphrases, and all testimony is presented in appropriate context.

BEFORE THE TRIAL

On September 9, 2015, Anita was issued a summons by a police officer, who came to her house and informed her that a case had been filed against her for criminal mischief for feeding water to thirsty pigs who were suffering in scorching hot temperatures inside filthy, crowded transport trucks.

This did not stop Anita and other Toronto Pig Save activists from standing up for the voiceless, vulnerable animals. They continued their actions of witness and mercy. On September 24, 2015, more than 100 people came to Fearmans Pork Inc. to provide love and kindness by giving water to the pigs before they went to slaughter.[5]

Anita had four preliminary hearings before her trial began.

On December 15, 2015, Anita spoke to reporters outside of the Burlington Courthouse at a preliminary hearing:

---

[5] Murphy, J. (2015, November 30). Canada woman faces 10 years in prison for giving pigs water on hot day. Retrieved February 17, 2020, from https://www.theguardian.com/world/2015/nov/30/canada-woman-10-years-prison-for-giving-pigs-water

I look forward to defending myself against these ridiculous charges. Compassion is not a crime. What I did was absolutely the right thing to do. Giving water to a thirsty animal is something any caring and reasonable person would do, without a moment's pause. Giving water to the thirsty is a duty we all share and is recognized universally as such in all religions and philosophies as a form of the Golden Rule.

The case was heard in the Ontario Superior Court of Justice and presided over by Justice David Harris.[6] Anita had two lawyers represent her: Gary Grill was the counsel and James Silver was the co-counsel.

The Counsel for the Crown was Harutyun Apel.

---

[6] Greenberg, J. (2017, March 30). The Anita Krajnc Trial: Compassion, the Public Interest, and the Case for Animal Personhood. Retrieved April 15, 2020, from http://ultravires.ca/2017/03/anita-krajnc-trial-compassion-public-interest-case-animal-personhood/

JEFFREY VELDJESGRAAF

Anita's trial began on August 24, 2016, at the Burlington Ontario Court of Justice on Plains Road. Jeffrey Veldjesgraaf, the truck driver who told Anita not to give water to the pigs and called the police, was the first witness.

"What was the approximate value of the hogs on that truck?"

Harutyun Apel, the counsel for the crown, examined Veldjesgraaf. The video of Anita feeding the thirsty pigs water was played for the court.

APEL: On the video, when you said you don't know what's the water, what did you mean by that?

VELDJESGRAAF: Just that. I have absolutely no idea what is in those bottles. It could be water; it could be anything.

APEL: And that day, were those hogs accepted by the plant?

VELDJESGRAAF: Yes.

APEL: What was the approximate value of the hogs on that truck?

VELDJESGRAAF: It—it fluctuates on a daily basis, but about $45,000.

THE COURT: It was $45,000?

VELDJESGRAAF: Yes.

"Who is your boss?"

James Silver cross-examined Jeffrey Veldjesgraaf.

SILVER: Did you have any issues with water being provided to the hogs?

VELDJESGRAAF: Yes, I did.

SILVER: Why?

VELDJESGRAAF: Those animals are in my care and control, and we—we're not allowed to let anybody feed an unknown substance to our animals while they're en route.

SILVER: And are those instructions you had received, sir, from somebody?

VELDJESGRAAF: Yeah. I would receive that from my boss.

7

SILVER: And your boss, who is your boss?

VELDJESGRAAF: Eric Van Boekel.

"That's over $23,000,000?"

SILVER: . . . You said the value of the hogs on this truck was about $45,000?

VELDJESGRAAF: Yeah. That's what I'm guessing. It's not my money, so I'm just—that was a guess.

SILVER: . . . You do that 52 weeks a year?

VELDJESGRAAF: Yes.

SILVER: That's over $23,000,000?

VELDJESGRAAF: Could be, yeah, if that's what it comes out to.

SILVER: You expressed your sentiments as clear as day?

VELDJESGRAAF: Yup.

SILVER: What you said was "They're not human, you dumb frickin' broad"?

VELDJESGRAAF: Right.

SILVER: When you came out of your truck, you said to her, "How do we know what's in the water?"

VELDJESGRAAF: Right.

SILVER: When you got out of the truck, you knew it was water, that's why you said how do we know what's in the water, right?

VELDJESGRAAF: It was a water bottle, yup.

SILVER: It's not how do we know what's in the water bottle, that's not what you said to her. What you said was, how do we know what's in the water?

VELDJESGRAAF: Okay. Yeah, it's a perfectly logical question, isn't it?

SILVER: And, shocked, she responded to you, water?

VELDJESGRAAF: Yup.

"They're messing with our livelihood"

SILVER: I'm going to suggest to you that you had no concern about what was in the water bottle, that's not what your concern was?

VELDJESGRAAF: Yeah, actually it is.

SILVER: She offered for you to take the bottle—take a sample of the water, test it if you wanted?

VELDJESGRAAF: She did?

SILVER: Do you want to hear the video again where she says that?

9

VELDJESGRAAF: I would love to take a sample of the water that she feeds all the time. Can we do that daily?

SILVER: We'll play the video again, please.

THE COURT: I'm not sure....

VELDJESGRAAF: I didn't hear her say that she would give us a sample of the water.

THE COURT: Okay.

SILVER: While we're getting that cued up, you didn't ask her for a sample?

VELDJESGRAAF: No.

SILVER: You didn't ask her for the bottle?

VELDJESGRAAF: I'm quite sure she wouldn't have given it to me.

SILVER: You didn't ask her for the bottle itself?

VELDJESGRAAF: No, I didn't. No.

SILVER: Can we watch it again, please.

. . . VIDEO PLAYED

SILVER: She invited you to test it—

VELDJESGRAAF: Yup.

SILVER: —she invited you to take a sample?

VELDJESGRAAF: Yup.

10

SILVER: You didn't take her up on that?

VELDJESGRAAF: . . . I didn't take her up on it because I didn't hear her.

SILVER: Because that was not your concern?

VELDJESGRAAF: My concern is what goes inside the trailer. I think I know what my concern is.

SILVER: I'm going to tell you what your concern is, sir. I appreciate that's what you're saying now. Your concern is what you told the police on July 26th, 2015. Your concern is this needs to be stopped, because they're messing with our livelihood?

VELDJESGRAAF: Yup.

SILVER: That's what you told the police on July 26th, 2015.

VELDJESGRAAF: That's right.

ERIC VAN BOEKEL

After Jeffrey Veldjesgraaf took the stand, Eric Van Boekel, his boss, the owner of Van Boekel Hog Farms, was examined by the counsel for the crown, Harutyun Apel.

"A great substantial amount of financial loss"

APEL: Sir, I understand that you are the owner of Van Boekel Hog Farms located in Mount Elgin, Ontario. Is that correct?

VAN BOEKEL: Yes. I am.

APEL: And I understand that you have a contract with Sofina Foods Incorporated, is that correct?

VAN BOEKEL: That is correct.

APEL: And the contract is in relation to you raising hogs and then transporting them to Fearmans Pork Incorporated, is that correct?

VAN BOEKEL: That is correct.

APEL: Where they are processed for human consumption?

VAN BOEKEL: That is correct.

APEL: On your farm, I understand you have staff, of course, correct?

VAN BOEKEL: That is correct.

APEL: Approximately how many?

VAN BOEKEL: Approximately 30 individuals work for us.

APEL: And you have livestock transporters, is that correct?

VAN BOEKEL: That is correct.

APEL: And one of them is Jeff, the gentleman that just testified, is that correct?

VAN BOEKEL: That is correct.

APEL: What concern did you have in your mind as to what Fearmans may do if outside contaminants go into the trailer?

VAN BOEKEL: With regards to Fearmans, my greatest fear is that if there's outside contaminants into my trailer load of pigs, that the load would be condemned. That would be my greatest fear for my own personal reasons. The other fear that I have is that contamination, if it's not caught early enough, could render the entire kill floor contaminated, could face liabilities upwards of a great substantial amount of financial loss.

"There's not sunlight"

James Silver questioned Van Boekel about the conditions in his farm.

———————

SILVER: Sun lights? Are there sun lights up in the ceiling or roof?

VAN BOEKEL: No. The—especially the sow barns, what happens with sows, and to maximize production again, is that they—the natural farrowing cycle of an animal goes on number of days—hours of daylight. So, when the sows are kept indoors, they—we monitor the lighting system so that they always have greater than 15 hours of sunlight—14 to 16 hours of sunlight is the maximum desired for farrowing and breeding sows.

SILVER: So during the winter how do you give them that amount of sunlight? There's not that much sun during the day.

VAN BOEKEL: . . . We use ultraviolet lights so that it mimics the sun.

SILVER: Okay. So, when I'm asking about sunlight, I'm talking about light that comes from the sun. I'm sorry.

VAN BOEKEL: You asked how we do the light.

14

SILVER: Okay.

VAN BOEKEL: And you asked if there was sunlight. I said no, there's not sunlight.

SILVER: Unless I've misunderstood what you're saying, it seems that you've replaced the provision of sunlight, per se, with the use of artificial lights to maximize the breeding cycle, I guess?

VAN BOEKEL: That's correct.

> *"Every pig in your facility has received*
> *antibiotics at some point in time?"*

SILVER: Do you administer antibiotics to the pigs?

VAN BOEKEL: We use it sparingly. We use antibiotics to maximize efficiency.

SILVER: What type of antibiotics do you use?

VAN BOEKEL: We'll use penicillin. We use a sulphur-based product. We'll use scour prevention products for the little pigs.

SILVER: You give all the little pigs certain antibiotic compounds?

VAN BOEKEL: Yes.

SILVER: Again, I'm going to ask you this. Tell me if I've misunderstood. It sounds like every pig gets antibiotics then?

VAN BOEKEL: At birth.

SILVER: Okay. So, every pig in your facility has received antibiotics at some point in time?

VAN BOEKEL: Yes.

SILVER: And the way that it's administered, is it injected, is it put in their food, how do they get it?

VAN BOEKEL: It's injected.

SILVER: And then aside from that initial administering of the antibiotics, is there a subsequent dosage which is provided?

VAN BOEKEL: No. What we'll do is because there's disease pressures through—that the pig will encounter later in life, we give them vaccines.

SILVER: And what type of vaccine is that?

VAN BOEKEL: The most prevalent right now is the circovirus vaccine.

SILVER: Okay.

VAN BOEKEL: Which came in from the deer.

SILVER: And when is that administered?

VAN BOEKEL: That's administered about day five, at weaning, day 21, and one more time at day 35.

SILVER: Do you know specifically how much antibiotics are administered during the year, on an annual basis?

VAN BOEKEL: Each little pig gets a quarter cc of penicillin mixed with 75 grams . . . so one cc in total of iron and penicillin mix.

"Do you use anaesthetic for this procedure?"

Mr. Silver questioned Mr. Van Boekel about the process where his workers cut the teeth and dock the tails of the pigs.

———————————

SILVER: You talked about teeth being cut and tails being cut. Is that a practice that your operation engages in?

VAN BOEKEL: They will use their incisor teeth against their litter mates to fight for teat space. They will scratch each other's faces. They will cause damage. Inadvertently, they will sometimes bite on the sow's teat, causing it to get infected. . . . Now, with the docking of the tails, in confined housing units, because pigs are territorial animals, they will bite each other's tails.

SILVER: And how do you clip the tails?

VAN BOEKEL: We use a different instrument, and we cut the tails.

SILVER: Like a pair of scissors type of thing?

VAN BOEKEL: No. It's a special instrument, and it crushes the tail, and it breaks the tail, and it's—I can liken it to a boy being circumcised. Nobody remembers, but it hurts like hell when it happens.

SILVER: Do you use anaesthetic for this procedure?

VAN BOEKEL: No, we don't.

SILVER: And do you use anaesthetic for the procedure that you use in order to cut the teeth?

VAN BOEKEL: No, we don't.

"Do you remember getting those letters?"

Silver asked Van Boekel about letters that were sent to him from Environment Canada, warning that he violated the *Federal Fisheries Act*.

SILVER: October 18th, 2006, a formal written warning letter was sent by registered mail to the company, and Eric Van Boekel, by Environment Canada inspector Brad Simpson, warning them that they had violated s. 36(3) of the *Federal Fisheries Act*, deleterious substance into a fish habitat. On October 24th, 2006, a compliance letter and compliance promotion information respecting manure

18

and environmental impact was also sent to Eric Van Boekel and his company. No charges were laid as a result of this incident. That's what the decision says. Do you remember getting those letters?

VAN BOEKEL: Yes.

SILVER: In 2006?

VAN BOEKEL: Yes.

"Eric Van Boekel was found guilty of that, yes"

Despite having been warned once, Eric Van Boekel and his company were charged for discharging deleterious substances into the Thames River and Sweets Creek in 2007.

On April 29, 2007, a farm worker attended the farm located on Braemar Sideroad, in the Township of East Zorra-Tavistock, after a pipe inside of the main barn at Van Boekel's Hog Farm burst, causing deleterious liquid to overflow from the barn and gush into the Thames River.

The farm worker who attended the spill did not notice the liquid had overflowed from the barn and entered the Thames River. On May 1, 2007, Jean Read, a nearby neighbour of Van Boekel Hog Farms, noticed the spill and complained about improper illegal spreading of pig manure at the same location.

On May 3, 2007, a Provincial Officer and two inspectors from Environment Canada went to Van Boekel Hog Farms and saw the water had been polluted and become murky and had entered Sweets Creek through a hidden tile drain without a catch basin underneath the creek bed.[7]

All of this may seem ironic given the suggestion that had been made about whether the water Anita Krajnc gave the pigs was contaminated.

After a break, Mr. Silver continued to question Mr. Van Boekel.

SILVER: Sir, you've had an opportunity over the break to review the factual underpinnings of your findings of guilt in relation to some Environmental Protection Act charges, correct?

VAN BOEKEL: That's correct.

SILVER: . . . And the incident, which was the discharge of contaminants into the natural environment and into water sources, those incidents occurred in 2007, your findings of guilt were in 2008, correct?

VAN BOEKEL: No. I believe it was 2012 or '13.

---

[7] Saxe, D. (2014, May 28). Huge Manure Spill Fines: $120,000 Plus Surcharge. Retrieved March 5, 2020, from https://www.lexisnexis.com/legalnewsroom/environmental/b/cleanaircleanwater/posts/huge-manure-spill-fines-120-000-plus-surcharge

SILVER: . . . You had a trial in that matter?

VAN BOEKEL: Yes, we did.

SILVER: You were found guilty of various charges?

VAN BOEKEL: Yes, we were.

SILVER: You ultimately appealed that?

VAN BOEKEL: Yes, we did.

SILVER: Some of the charges were—some of the convictions were overturned?

VAN BOEKEL: That's correct.

SILVER: You ultimately came back and pled guilty to a selection of various charges?

VAN BOEKEL: Yes, we did.

SILVER: And ultimately you were fined?

VAN BOEKEL: Yes, we were.

SILVER: The circumstances surrounding that—those findings of guilt had to do with the discharge of—really what we've called this compendium of substances into the environment?

VAN BOEKEL: Yes.

SILVER: And by compendium of substances, I'm talking about the manure, urine, that kind of sludge of stuff that is held in those waste containment units?

VAN BOEKEL: Yes.

SILVER: . . . Amounts of this substance made their way into Sweet River?

VAN BOEKEL: Sweet River Creek, yes.

SILVER: . . . there was a serious fill of a large quantity of what was believed to be pig manure running from the east and north sides of the barn, down the barn hill, across the driveway, around their drinking water well, surrounding their house, and continuing onto and into the Thames River? That's what you were found guilty of, right?

VAN BOEKEL: Eric Van Boekel was found guilty of that, yes.

SILVER: Sorry, you're Eric Van Boekel, right?

VAN BOEKEL: Yes.

SILVER: Okay. So, when you say Eric Van Boekel was found guilty, you were found guilty?

VAN BOEKEL: Yes.

> *"You said you regret it very much.*
> *You now regret it very much?"*

SILVER: Okay. The Environmental Protection Agency conducted this investigation?

VAN BOEKEL: Yes.

SILVER: It was not only you who were found guilty, but also Van Boekel Hog Farms, right?

VAN BOEKEL: Yes. . . . It was an unfortunate incident, we regret it very much. It was a low point in the company's time. We made restitution with the courts. We've paid our fines. I have, as I described earlier, a large operation. I have some unique individuals, very good men that work for us, hardworking, caring, but as things happen, they made a few mistakes. Because I am the director and the owner of Van Boekel Hog Farms, I lay it on my shoulders and take full responsibility. We have made changes to ensure that it doesn't happen again. We all have to learn from our mistakes. That was a mistake in our operation.

SILVER: You said you regret it very much. You now regret it very much?

VAN BOEKEL: No, I regretted it then. That's why we pled guilty.

SILVER: . . . And yet you had a trial.

VAN BOEKEL: . . . Yes. But that doesn't mean that we didn't regret it happened.

SILVER: And then you had an appeal?

VAN BOEKEL: That's correct.

SILVER: And then you pled guilty?

VAN BOEKEL: And then we pled guilty. Exactly.

SILVER: This is a final guilty plea . . .

VAN BOEKEL: All regrettable offences.

SILVER: Final guilty plea some five years after the regrettable incident itself?

VAN BOEKEL: Yes.

"Your concern is food safety?"

Having established that Eric Van Boekel and his company keep pigs without daylight, cut their teeth and their tails, and dose them regularly with antibiotics, and that Van Boekel and his company had been fined for polluting the environment and had fought the charge as far as they could before pleading guilty, James Silver had just a few more questions for Eric Van Boekel.

VAN BOEKEL: I'm not anticipating any problems. I have no problems at all with Anita Kromwell [sic] or the save pig program when they protest in a legal and responsible manner. When they are on the side of the road and not interfering with my property, I bless

24

them. They—our nation, Canada, is founded on the principles that people can protest for their rights.

SILVER: Your concern is food safety?

VAN BOEKEL: Food safety, animal welfare, it is—it is not more than once or twice, countless instances where animal activists in the name of welfare have caused hundreds, thousands, millions of dollars worth of damages.

SILVER: It hasn't cost you a penny?

VAN BOEKEL: Not yet.

SILVER: In years?

VAN BOEKEL: Not yet.

DR. DAVID JENKINS

The defence called expert witnesses to support Anita's case.

Dr. David Jenkins is an expert on human nutrition, and he was permitted to give his expert opinion regarding the nutritional health impact on humans from the consumption of various foods and the scientific relationship between animal products and chronic diseases, specifically diabetes, heart disease, and cancer.

Dr. Jenkins was educated at Oxford University and is currently a Professor in both the Departments of Nutritional Sciences and of Medicine at the University of Toronto, a Staff Physician in the Division of Endocrinology and Metabolism, the Director of the Clinical Nutrition and Risk Factor Modification Center, and a Scientist in the Li Ka Shing Knowledge Institute of St. Michael's Hospital. He has for decades been one of the world's leading researchers in the relation of diet to chronic disease.

"Diabetes, cardiovascular disease, and cancer"

Anita's co-counsel, James Silver, questioned Dr. Jenkins.

SILVER: I remember as a kid, you know, my parents would say things like "Eat your meat if you want to get strong and big." Is that wrong?

JENKINS: No, that's absolutely so. And so, if you want to get your cancer to grow big and strong, then that's perhaps the way you can do it. And that is the problem, and you're raising a very important point. The things that nourish us, nourish our healthy cells, remember, are also the things that nourish your cancer cell. So, if you have an increased nourishment of any kind, that's why people who are overweight, people who are diabetic, for example—diabetes increases your risk of getting a cancer by 30 or, more, 40 percent.

SILVER: Can you say whether there's been any link in the literature between increased consumption of red meat and increased incidents of type 2 diabetes?

JENKINS: Yes, there has—diabetes, cardiovascular disease, and cancer. The large-cohort studies have shown that there is a link between the consumption of meat, and in fact of animal products in general—animal protein, which is largely going to be red meat protein—and diabetes and cardiovascular disease. . . .

SILVER: Now, as far as the WHO goes, were they able to quantify the increased risk of cancer based on the consumption of processed meats?

JENKINS: There are no numbers given, but as I say, the numbers that we're talking about are 3 percent of cancers related to meat consumption and probably 18 to smoking, to give you a sort of a reference. And I think that if you want to go to other situations, other than cancer, then I think one looks at the diabetes associations and one looks at the heart associations for their advice. And I think on that point you can make another statement, and that is that since 2012, the Canadian Cardiovascular Society, which sets the guidelines for the treatment of cardiovascular disease in Canada, on its nutrition panel came to the conclusion that there are three diets that we should be recommending. One diet is the Mediterranean diet, which is a lower-meat diet, traditionally. The second was the DASH diet, which specifically states reduce meat intake. And the other was the dietary portfolio, which is in fact a vegan diet. And those were their three diets, and they've been in the latest guidelines that are now being put together this year. It's the same pattern of diets with a greater emphasis on plant proteins as being the replacement. . . . The Canadian Cancer Society also says, as you'd expect . . . reduce meat intake. . . . The Diabetes Association has placed the vegetarian diet as one of its key diets for the treatment of diabetes.

So, I think what you'll see—and I think this is the relevant theme—internationally, one's seeing a move away from a meat consumption diet towards a more plant-based diet. The mantra is eat more fruit, vegetables, whole-grain cereals, legumes, nuts and seeds. That is the mantra and it's a recognized mantra, nationally and internationally. But one's seeing it in Canada for cancer, one's seeing it for diabetes, one's seeing it for cardiovascular disease.

Another detail that Dr. Jenkins didn't point out is that bacon has been classified by the World Health Organization as a carcinogen, in the same category as cigarettes and asbestos.[8]

"They become antibiotic resistant, as you can imagine"

SILVER: Why is antibiotic use in a factory farm setting of significance to you as a nutritionist?

---

[8] "World Health Organization Says Processed Meat Causes Cancer." *American Cancer Society*, www.cancer.org/latest-news/world-health-organization-says-processed-meat-causes-cancer.html.

JENKINS: I think it's not necessary as a nutritionist, but from a point of view of clinical medicine, factory farming and the use of antibiotics in factory farm animals is one of our concerns in terms of the production of resistant bacteria. They become antibiotic resistant, as you can imagine; whenever you give antibiotics you run the risk of antibiotic resistance developing in the bacteria.

What is antibiotic resistance?

Dr. Alexander Fleming discovered penicillin, the world's first true antibiotic, in 1928. Doctors could do little for people with infections before Fleming discovered penicillin.[9] There was no treatment for infections such as pneumonia, scarlet fever, and Lyme disease.[10,11] Hospitals were filled with people with blood poisoning contracted from a cut or scratch.

In 1945, Fleming was jointly awarded the Nobel Prize in Physiology or Medicine with Ernst Boris Chain and Sir

---

[9] Discovery and Development of Penicillin... (n.d.) Retrieved March 4, 2020, from https://www.acs.org/content/acs/en/education/whatischemistry/landmarks/flemingpenicillin.html

[10] Alexander Fleming Discovery and Development of Penicillin ... (n.d.). Retrieved March 4, 2020, from https://www.acs.org/content/acs/en/education/whatischemistry/landmarks/flemingpenicillin.html

[11] What Does Penicillin Treat? (2016, January 25). Retrieved March 4, 2020, from https://www.newhealthadvisor.org/What-Does-Penicillin-Treat.html

Howard Walter Florey "for the discovery of penicillin and its curative effect in various infectious diseases."[12]

Fleming warned the public about the dangers of penicillin resistance in 1946: "The thoughtless person playing with penicillin treatment is morally responsible for the death of the man who succumbs to infection with the penicillin-resistant organism. I hope this evil can be averted."[13]

Animals can carry bacteria that can grow and survive inside of their intestines. When the animals are given antibiotics, some bacteria that are not killed by it will continue to grow and, in the absence of competition from the more susceptible bacteria, will become dominant, so that the antibiotics will no longer be effective. The animals will then carry and spread these resistant bacteria. When animals are trucked to slaughter and killed in slaughterhouses, their bodies can become contaminated. Animal feces can also contaminate fruits and vegetables when resistant bacteria enters the environment, through irrigation water or fertilizers.

Antibiotic-resistant infections can increase the probability that a person will be hospitalized. It may take a person

---

[12] The Nobel Prize in Physiology or Medicine 1945. (n.d.). Retrieved March 4, 2020, from https://www.nobelprize.org/prizes/medicine/1945/ceremony-speech/

[13] Penicillin's finder assays its future. New York Times, June 26, 1945, p. 21.

longer to heal, and people are dying from infections that were once treatable with antibiotics.[14]

According to the World Health Organization, antibiotic resistance is one of the biggest threats to food security and global health today. A growing list of infections such as pneumonia, tuberculosis, blood poisoning, and gonorrhea are becoming harder, and sometimes impossible, to treat as antibiotics become less effective.[15]

Each year in the world, as a direct result of these antibiotic-resistant infections, at least 700,000 people die, and the death toll from antibiotic resistance is expected to rise in the future. It is estimated by the year 2050 more people will die from antibiotic-resistant infections (ten million) than from cancer (eight million).[16]

[14] Antibiotic Resistance and NARMS Surveillance. (2019, November 21). Retrieved March 4, 2020, from https://www.cdc.gov/narms/faq.html

[15] Antibiotic resistance. (n.d.). Retrieved from https://www.who.int/news-room/fact-sheets/detail/antibiotic-resistance

[16] McKenna, M. (n.d.). What do we do when antibiotics don't work any more? Retrieved March 4, 2020, from https://www.ted.com/talks/maryn_mckenna_what_do_we_do_when_antibiotics_don_t_work_any_more

DR. TONY WEIS

Dr. Tony Weis has a PhD in Geography from Queen's University and is Associate Professor of Geography at Western University in London, Ontario, where he has taught courses on world agriculture and food systems, global environmental change, biodiversity laws, and climate change. He has written two books: *The Global Food Economy: The Battle for the Future of Farming* and *The Ecological Hoofprint: The Global Burden of Industrial Livestock.*

Dr. Weis was sworn in as an expert in geography and environmental studies. He gave his opinion evidence on the historical transformations of agriculture and environmental impacts of industrial livestock productions, and the social impacts of the globalization of livestock production.

"120 billion animals will be killed for food every year"

Weis was examined by Anita's counsel, Gary Grill.

GRILL: You indicated in your expert report at page 2, at the bottom of paragraph 1, "rising meat production and consumption is a momentous and destructive aspect of world agriculture." You go on to say, "although Canada ranks amongst the highest per capita producers and consumers of meat in the world, few Canadians give any thought to where their meat comes from, the environmental impacts of its production, the fact a large share of Canada's best arable land is devoted to feed crops along with large amounts of energy, water and other inputs where the conditions under which livestock populations live and die."

WEIS: Yeah, I think that's a very important statement that captures a lot of what I was trying to convey. But on, again, on a world scale, the average person was consuming 23 kilograms of meat per year in 1960. Today it's around 43. So, not quite a doubling, and what's also important about that average is that the human population in that time has grown from about 3 billion to 7.3 billion today. So, you have a period of more than a doubling of the human population and the average person on earth consuming almost twice as much meat on an annual basis, and that trajectory the FAO projects to continue growing. The FAO's projection is that by 2050, in a world of 9 to 10 billion people, the average person will be consuming over 50 kilograms of meat per year.

GRILL: If, if the average person in the world is consuming now 43 and it's projected to go up to 50—

WEIS: Over 50.

GRILL: —over 50, what is the average North American—or the average Canadian, or American, consuming?

WEIS: Okay, that's another big point I stress with meatification, is that it's an incredibly unequal phenomenon on a world scale. The average Canadian consumes over 100 kilograms of meat per year, the average American closer to 120 kilograms of meat per year, and the average person in sub-Saharan Africa under 20 kilograms of meat per year, the average person in South Asia under 10. So, there's an incredible inequality in terms of the volume of meat consumed on an annual basis, and tied to that is inequality in grain products.

GRILL: You make the point that 70 billion animals worldwide are killed for food every single year.

WEIS: Yeah, the number's over 70 now, and again, to historicize that, in 1960 that was somewhere between 7 and 8 billion, and now we're over 70 billion, and I have a recent paper that's called "Towards 120 Billion". . . if dietary change continues as projected that I was indicating, the average person in 2050 consuming over 50 kilograms of meat per year in a world of 9 to 10 billion, the expectation is that 120 billion animals will be killed for food every year.

"Humans are driving the sixth extinction spasm"

GRILL: Tell me a little bit about biodiversity. What do you mean by that?

WEIS: "Biodiversity" is a broad term that encapsulates the range of species that exist on earth, and it is now widely recognized that we are in the midst of the sixth extinction spasm. In the course of evolutionary history there have been five other extinction spasms where [there was a] very rapid loss of a significant share of all of the species existing on earth at that time. [It is] very well recognized that humans are driving the sixth extinction spasm, meaning that species are being driven off this planet.

GRILL: Are you talking about animal and plant species?

WEIS: Yeah, animal and plant species. If you just look at mammals, say, about a quarter of all mammals on earth are projected to be extinct in the coming century; a quarter, over a tenth of all birds, about 40 percent of all amphibians. Now, there's a range of different estimates, and again, by that I don't mean to in any way suggest that it's inevitable, but that's what the trajectory, the best conservation science is saying: if you look at rates of endangerment, threatened species, if the current trajectory continues we are looking at these phenomenal losses of fish, birds, mammals, and all kinds of other smaller species. . . . The heart of my book *The Ecological Hoofprint: The Global Burden of Industrial Livestock* was to say this is a momentous environmental issue that bears on so many of the world's urgent environmental problems, from climate change to biodiversity loss, and it is also tied to incredible human inequalities.

DR. ARMAITI KHORSHED MAY

Dr. Armaiti Khorshed May is an expert in veterinary medicine and animal welfare. In 2001, she graduated from the University of California at Berkeley with a bachelor of science degree in bioresource sciences, and in 2005 from UC Davis School of Veterinary Medicine as a doctor of veterinary medicine. In 2007, Dr. May became a certified veterinary acupuncturist, and was certified in animal chiropractic in 2015. She had been a practising veterinarian for over 11 years at the time of the trial, eight of which as a house-call vet and before that in veterinary emergency hospitals. She has also done volunteer work on Native American reservations.

Dr. May was called to give opinion evidence regarding pig physiology and responses to stimuli and environmental conditions.

"Many of them were suffering from overheating"

James Silver questioned Dr. May.

SILVER: Now, you've been qualified to provide an expert opinion on this topic, of pig physiology and responses to stimuli in environmental conditions. In preparation to provide this opinion, have you reviewed any material?

A. I did.

SILVER: What material did you review?

MAY: The video that was provided to me, which showed the man who came out of the truck and interacted with the woman who was giving water to the pig in the truck.

SILVER: And it's been conceded, and it's an accepted fact that on that day, although we don't know that particular exact time, that day the ambient midday temperature was 26.1 degrees Celsius, which is 79 degrees Fahrenheit, and the humidity level was 61 percent?

MAY: That's right.

SILVER: Let me ask you about the pig that we see receiving water in that video. Have you been able to formulate any opinions about that pig's condition?

MAY: Yes. The pig very eagerly and rapidly accepted the water, which, amongst other considerations, is an indicator of that pig's need for the water. There were a number of pigs in that truck who demonstrated heavy panting, which also was one of the signs of overheating, along with an increased respiratory rate of about 180 breaths per minute, which is greatly exceeding the normal respiratory rate of eight to 18 breaths per minute.

SILVER: When you say a number of pigs, did you notice this particular pig?

MAY: The video was hard to say which pig was the same pig who was doing the panting from the back and forth, but from looking at the pigs in the truck, it does appear that many of them were suffering from overheating.

SILVER: And you draw that conclusion from?

MAY: The fact that their respiratory rate was elevated, they were open-mouth panting, and at least in one instance, a pig had some foaming at the mouth, which is also suggestive of overheating and heat distress.

SILVER: Are you able to indicate to the court how a pig feels when it's overheating or suffering from heat distress?

MAY: Pigs, unlike other mammals, are not able to sweat, and that means that they have to rely on other mechanisms to cool their bodies when there's excessive ambient heat. The humidity adds to the feeling of a warm day because it actually increases the effect of the heat in their bodies, and that makes them have to rely on panting as a primary mode of cooling if they don't have water given either as misting, or spraying on them, or allowing them to drink the water to alleviate the excessive heat they're experiencing.

SILVER: Are you able to render an opinion about the effect that would be had as a result of water being offered?

MAY: Yes. Water being offered in this instance provided relief of discomfort, alleviation of suffering, and, to some measure, alleviation of the dehydration these pigs were experiencing.

DR. LORI MARINO

Dr. Lori Marino is an expert in neuroscience, animal behaviour, and biopsychology. She holds a bachelor's degree from New York University in psychology and biology, an MA in experimental psychology from the University of Ohio, and a PhD in biopsychology from the State University of New York at Albany.

"It sells pigs short to say that they're like a human toddler"

Dr. Marino was asked a series of questions by Anita's counsel, Gary Grill.

GRILL: How many animals are as complex as pigs?

MARINO: There are not that many. When you really look at the totality of how complex they are in their cognitive abilities, their memories, their social structures, their emotions, their personalities, we're talking about pigs being at the very least as emotionally complex as dogs and, in many cases, as psychologically complex as primates. . . . We see a lot of behaviours in chimpanzees that we see in a very few number of species, including the great apes.

GRILL: . . . I've heard suggestion pigs have the cognitive abilities of a two-year-old, four-year-old human child; is that accurate?

MARINO: . . . There is the fact that there are some things that a two- to four-year-old human can do that pigs can also do. But I think it sells pigs short to say that they're like a human toddler; I think they're more sophisticated than that.

Gary Grill asked Dr. Marino about a study where pigs were determined to have the cognitive ability to play video games.

MARINO: Yes, this was a landmark study done by Dr. Candace Croney at Purdue. The study is important because when you play video games you have to have a capacity called "self-agency"; that has to do with self-awareness. You have to know, when you do an action, that that impacts something else. So, when you play a video game, you're using a joystick and you're moving a cursor on a screen, that's the basis of it, and in order to really understand that you're doing that, you have to

43

have a sense of self. And what Dr. Croney showed is that yes, indeed, when you ask pigs to move a cursor on a screen to a target, they can use a joy stick to do that. And that, again, is something that is a very rare capacity in the animal kingdom.

"Yes, of course they experience fear"

Gary Grill asked Dr. Marino if pigs feel pain the same way as humans do.

GRILL: Do they feel and experience pain in the same ways?

MARINO: Yes, they do, and the reason I can tell you that is from inferential neuroanatomy, which is a fancy way of saying if I look at the brain of the pig, I see the same parts in it that are related to not only pain reception, but other kinds of aspects of psychology and suffering in human brains. They have all the brain parts that you need to experience distress, certainly pain.

GRILL: If you are repeatedly being physically injured on a daily basis, do pigs suffer psychologically as a result of the repeated, and I'm just going to use the word abuse, on a regular basis?

MARINO: Yes.

GRILL: So they experience fear.

MARINO: Yes, of course they experience fear.

GRILL: They feel anxiety.

MARINO: Yes, they do.

GRILL: Oftentimes and during the course of this trial we've heard the word "sentient" being used. Are pigs sentient?

MARINO: Yes, they are sentient.

GRILL: What does that mean at its most basic level?

MARINO: . . . At a most basic level it's just the ability to feel. It's the ability to feel both good and bad depending upon what has happened to you.

GRILL: And in positive environments, do they experience feelings of joy and happiness?

MARINO: Yes.

A video was played in the courtroom displaying pigs who were forced to go into gas chambers to be knocked out before they were slaughtered.

Before pigs are slaughtered, they are lowered into gas chambers, stunned and forced to inhale carbon dioxide in high concentrations. When the pigs are lowered into the gas chambers, they gasp for air, as each breath of CO_2 that they inhale burns them alive from the inside out.[17] The CEO of

---

[17] Zampa, Matthew. "There's Nothing 'Humane' About Killing Pigs in Gas Chambers." *There's Nothing "Humane" About Killing Pigs in Gas Chambers*, 17 Dec. 2019, sentientmedia.org/nothing-humane-about-killing-pigs-gas-chambers/.

Compassion in World Farming explains, "The gas acidifies eyes, nostrils, mouths, and lungs."[18]

Pigs can regain consciousness for a variety of reasons; they are overloaded into gas chambers and they do not always inhale adequate CO_2 to remain fully unconscious before they are slaughtered. Pigs recover consciousness some of the time before they are killed.[19] They are not given any anesthesia.

"Are pigs persons?"

The video of Anita Krajnc feeding the thirsty pigs water was played in the courtroom. Gary Grill continued to question Dr. Marino.

GRILL: Doctor, are you able to give an opinion about the state of the pigs in this transport truck?

MARINO: Well, just from what I can see they are panting, they are frothing at the mouth, and those are signs of the inability to regulate temperature; they're probably overheated and dehydrated. In terms of their body

---

[18] Philip Lymbery. "IS GAS KILLING THE PIG INDUSTRY'S DARKEST SECRET?" *Philip Lymbery*, 4 Nov. 2019, philiplymbery.com/gas-killing-is-pig-industrys-darkest-secret/#.

[19] Zampa, Matthew. "There's Nothing 'Humane' About Killing Pigs in Gas Chambers." *There's Nothing "Humane" About Killing Pigs in Gas Chambers*, 17 Dec. 2019, sentientmedia.org/nothing-humane-about-killing-pigs-gas-chambers/.

language and some of the squeals that I hear, it looks like they're also obviously in psychological distress.

GRILL: Thank you. Are you able to define the term "personhood"? . . . What does it mean to be a person?

MARINO: What it means to be a person is not equivalent to being human. A person, even under the law, is someone who is autonomous, has the capacity to be self-aware and has the ability to understand that they have a life that they are going to be leading, a future. The common law says that practical autonomy is equivalent to personhood.

GRILL: By that definition, are pigs persons?

MARINO: Yes, they are persons.

"I think it's clear that pigs are being
tortured in a factory farm setting"

GRILL: . . . An accepted definition of torture is as follows: "the action or practice of inflicting severe pain on someone as punishments, or to force them to do or say something, or for the pleasure of the person inflicting the pain." Based on that definition, can you give an opinion as to whether or not pigs are being tortured in a factory farm setting?

MARINO: Based upon that definition, I think it's clear that pigs are being tortured in a factory farm setting.

DR. ANITA KRAJNC

Anita Krajnc was called to testify in her own defence.

Dr. Anita Krajnc holds a PhD in Political Science from the University of Toronto. She is the founder of Toronto Pig Save as well as the Save Movement, an international animal rights organization dedicated to bearing witness to vulnerable animals being transported to their deaths in crowded transport trucks.[20] [21]

Toronto Pig Save began when Anita Krajnc was walking her late dog Mr. Bean. Together, they saw animals going to slaughter in confined transport trucks in Toronto.

Anita began holding weekly vigils using a non-violent, love-based approach, and bearing witness to animals being trucked to their death, first with a small group of activists, and now with Toronto Pig Save. The Save Movement has expanded to over seven hundred groups around the globe.

---

[20] The Save Movement. (n.d.). Retrieved March 5, 2020, from https://thesavemovement.org/animal-save-movement/

[21] Toronto Pig Save. (n.d.). Retrieved March 5, 2020, from https://torontopigsave.org/

Toronto Pig Save holds vigils at the Fearmans Pork Inc. Slaughterhouse in Burlington on a weekly basis. The incident on June 15, 2015, where Jeffrey Veldjesgraaf confronted Anita, was a Toronto Pig Save vigil.

On October 3, 2016, Anita took the stand in her trial.

"If people don't see the injustice then
they're not going to think about it"

Gary Grill asked Anita about the importance of being there for the victims and bearing witness at slaughterhouses.

GRILL: Why is that so important, to actually witness it for yourself? Why is it so important for everybody to witness it?

KRAJNC: There's nothing like witnessing an injustice first-hand, it can change you forever. That's what happened to me: I was already a vegetarian and then I became a vegan; I was an activist. But when I first bore witness it changed me forever, it changed the priorities in my life. . . . And you know, if you look at historically why groups have promoted the idea of bearing witness, it's because there's this mass disconnect in society; if people don't see the injustice then they're not going to think about it, they're not going to work to change the world.

James Silver continued the questions.

SILVER: Can you tell me what you mean by a mass disconnect?

KRAJNC: When you go to the supermarket, you don't see the animal's eyes, you don't hear their cries. You know, when a sow's piglets are taken away from her or when a calf is taken away from a mother cow, you don't hear their cries for weeks. In a supermarket you just see, you know, cellophane-wrapped meat and, you know, I used to participate in that and I was disconnected so what we're trying to do is to show the truth. . . .

"They're just like people"

KRAJNC: I want to go through eight points about who pigs are. . . . At the pig preserve . . . in Jamestown, Tennessee, it's 100 acres and there's 120 pigs that roam freely in this area. What do they eat? They eat basically nuts, fruits, and grasses; so, they eat fallen acorns, hickory nuts, walnuts, wild squash and pumpkin, blackberries. They root and they eat different types of grasses; some pigs like some types of grasses, some others. They're just like people, they have different tastes.

50

How far do they travel in a day? . . . in a factory farm sows can't even turn around, they can't even move forward. They're stuck in a gestation crate and we know that in factory farms they're crammed; thousands of pigs are crammed in warehouses. How far do they travel at the pig preserve? They're roaming animals, they're very active. So, a farm pig that's fallen off a truck and has been rescued and is at the pig preserve will travel five to ten acres. Some of the young pigs will travel the entire perimeter, which is like 50 acres a day. . . .

You can imagine in a factory farm, like in a slave system, you have no control over your family. Your babies are taken away from you. That's what happens in factory farms. At the pig preserve, pigs form their own social groups. The groups will be two to 12 pigs; they form groups based on their disposition and personality. Not based on how big they are, their age or the type of pig they are; there'll be feral-type pigs forming groups with farm pigs.

What kind of language do pigs have? Richard Hoyle, who set up this pig preserve—he's an ex-marine, an ex-firefighter. . . . And he says—well, I heard a lot of the pig sounds, but they're about different vocalizations. But when you count facial expressions, like whether they open their mouths and show their tusks or their body posture, that leads to about 100 to 150 different types of communications.

What kind of rituals do pigs have? Pig societies are matriarchal, with the sow at the helm and they're by and large very cooperative; they have minor skirmishes that have more to do with who's in charge, but basically, they get along and cooperate. When a pig dies, pigs will go one by one to the dead pig and sniff the pig and talk to the pig and try to lift the pig up. In factory farms and at the slaughterhouse, sows can't protect their young; they're just taken away from them. And at the slaughterhouse, pigs have to watch other pigs being killed. There's no ritual for deaths, like we have funerals when people die and we give our respects. The slaughterhouse is a death camp. . . .

"In a factory farm they're covered in their excrement"

At the pig preserve, pigs enjoy mud baths. One way for them to keep cool is to go into the mud bath and also roam in the mud and it also keeps the insects away. In a factory farm they're covered in their excrement. Pigs are known as the cleanest animals, and at the pig preserve, they smell like maple syrup or like grass; at factory farms they smell like feces and it's just—it's horrific.

How long do pigs live? Well, we learned from Mr. Boekel that pigs are sent to market at four to six months; they're babies. And sows are sent to slaughter at three to four years after being constantly impregnated or, you know, artificially inseminated. At the pig preserve, pigs live 17, 18, 19 years . . . and potbellies will live in to their early 20s. So, basically, their lives are stolen from them in our animal agricultural system. . . .

And pigs have huge personalities, they're incredibly intelligent. They're affectionate, they're tender-hearted, they're just like people. Well, I think they're better than people, they're more noble; they don't have some of the negative sort of evil attributes that humans often express. They're curious, they're playful. I mean, they're just like dogs.

"You don't hear their cries"

SILVER: Why do you think people are still eating animals, eating pigs?

KRAJNC: I used to enjoy bacon, I used to enjoy pork as a teenager. Even a pig roast, I saw the pig and I didn't think at all of the pigs. So there's a mass disconnect and that's what has to change. And that's what our group is trying to do: it's bearing witness.

GRILL: The first video we saw was in 2011. Have you been conducting vigils and giving water to pigs since then?

KRAJNC: No, we only started giving water to pigs about three years ago, and we've been doing vigils for five years, so that's about 1,000 vigils since we started.

GRILL: Why did you give water to that pig?

KRAJNC: I gave water to the pigs before me because they were foaming at the mouth, they were thirsty.

GRILL: And you indicated that you were videotaping it. Why videotape it?

KRAJNC: At our vigils we always have a dual purpose. We're there for the animals before us, and because we can't save them and they go to slaughter, we have a bigger mission: we take videos so the whole world can see the animal victims. We want people to think about the 70 billion land animals that go to slaughter every year and the trillion of sea creatures that are killed each year. We also want people to think about the environmental impacts of animal agriculture. And we want people to think about their health and also their soul. . . . When you go to the supermarket, you don't see the animal's eyes, you don't hear their cries.

GRILL: Ms. Krajnc, if I understand correctly, the gas chamber doesn't kill the pig.

54

KRAJNC: That's right, the gas chamber—the carbon dioxide gas chamber is meant to stun the pig because they don't want to kill the pig because they want to hang the pig upside down with the heart still beating to pump the blood out once they slash their throat. Even the industry admits that two percent of the pigs are fully conscious awake when they're in the scalding tank. So, they're thrown into boiling water to loosen their hairs. A lot of the parts in the slaughterhouse they are just to loosen the hairs of the pigs.

"There's major water pollution
associated with animal agriculture"

Gary Grill questioned Anita about the impact of the animal agriculture industry and climate change.

GRILL: What does our environment have to do with eating meat?

KRAJNC: . . . Animal agriculture's widely known as polluting the waterways, with large feedlots and factory farms.

GRILL: You mentioned ocean dead zones, I thought, and you mentioned water pollution. Can you comment on those?

KRAJNC: Well, I can talk a bit about water pollution. I had published a paper when I was in political science, on cuts to the Ministry of Environment, right before Walkerton hit. I published this paper in 2000 and in May 2000 the Walkerton crisis hit. That was an *E. coli* outbreak where seven people died and thousands of people became ill from gastrointestinal infections and bloody diarrhea. And that contamination of water stemmed from a cow factory farm. So, it's widely known that there's major water pollution associated with animal agriculture.

"I'd like a stranger to give me water"

Harutyun Apel, the council for the crown, cross-examined Anita.

APEL: So, you think, in your mind, I'm just talking about in your mind, whether it was water, watermelons, you— anybody on the street—could just throw whatever they want in to that truck and let the pigs eat or drink it; no issue with that.

KRAJNC: It's the same as when people give sandwiches to homeless people. You don't say, "Is there—is that sandwich contaminated?" This is an act of charity. When we give water it's a small of charity, when we give water to thirsty animals—this is something that people have done for thousands of years, and I went through the history of bearing witness and giving water to thirsty animals.

APEL: You gave the example of you go in to a parking lot, it's a hot day, the windows are up and a dog is in the car, correct?

KRAJNC: Yes.

APEL: And you would say you wouldn't just walk by that, correct?

KRAJNC: That's right.

APEL: . . . So, is that your golden rule: if someone's thirsty you give them water.

KRAJNC: You should treat others as you'd like to be treated. If I was thirsty, I would want you to give me water; I'd like a stranger to give me water.

THE VERDICT

Compassion is not a crime

On May 3, 2017, nearly two years after Anita had fed the thirsty pigs water, Judge Harris found Anita Krajnc not guilty of mischief for giving water to pigs outside of Fearmans Pork Slaughterhouse Inc.

It was clear Krajnc was giving the pigs water, Harris said, and not an "unknown liquid" as police initially alleged. And the pigs were slaughtered anyway, which means she didn't obstruct their "lawful use."[22]

The decision was heard across the world. It helped build momentum for the fastest-growing social justice movement of our time.

No slaughter is humane

We don't all take the kind of compassionate action Anita

---

[22] Craggs, Samantha. "Animal Rights Activist Who Gave Pigs Water Found Not Guilty of Mischief | CBC News." *CBCnews*, CBC/Radio Canada, 4 May 2017, www.cbc.ca/news/canada/hamilton/pig-trial-verdict-1.4098046.

Krajnc did —and still does—but we all want to see ourselves as compassionate people. We all want to see ourselves as humane.

And many of us believe that we can eat meat and still be compassionate and humane, using "ethical" sources of meat from animals that have been handled in "humane" ways.

But, as Dr. Weis showed us, raising animals for killing and eating is also killing and eating the planet.

And no slaughter is humane.

All slaughter involves taking the life of an innocent sentient individual who did not want to lose his or her life for the sake of someone else's appetites. The most "humane" slaughter involves fear, suffering, and distress. Every time I have been to the slaughterhouses, I have witnessed animals being tortured and suffering in terror.

All use of animals is morally wrong. Non-human animals' lives are their right. Exploiting and killing them is violent and unjust, no matter how it is done. Humane or ethical killing is impossible. Sentient individuals do not want to die. Do you believe that an individual with culture and feelings should be raised in a factory, hauled in a filthy and crowded truck, and then violently killed in front of other members of their species just so a human can eat his or her skin and muscles?

One of the greatest books I ever read is Peter Buffett's *Life is What You Make It: Find Your Own Path to Fulfillment*. He shares some great wisdom:

> No one gets to choose his parents or to have a say in the circumstances of her birth. A life may begin in a snug and comfortable bedroom in an American suburb or on a straw mat in a mud hut in West Africa. The parents could be residents of a Park Avenue penthouse, or homeless people barely surviving in a public park. They might be healthy, or they might be infected with HIV. They could be athletes and scholars, or crack addicts and criminals. They could be partners in a committed couple for whom parenthood will be one of life's high points, or they could be virtual strangers out on a date, completely indifferent to the consequences of their actions.[23]

Or they could be living in a factory or concentration camp, treated as nothing more than... animals.

"Sometimes we must interfere"

I take inspiration from Dr. Alex Hershaft.

---

[23] Buffett, P. (2010). *Life Is What You Make It*. New York: Three Rivers Press. p. 27.

Dr. Hershaft was born in Warsaw, Poland, shortly before World War II began. When he was a child the Nazi army invaded Poland and ordered the Jewish people to move into the Jewish ghetto area of Warsaw. When its residents started being taken to the death camp at Treblinka, his family managed to escape, but his father was caught and murdered.

In 1972, Hershaft began working at an environmental consulting firm, and he was sent to a slaughterhouse in the Midwest United States. At the slaughterhouse, he saw piles of hoofs, skins, hearts, livers, and skulls. Hershaft tried to tell himself they were just animals. But he remembered the branding and tattooing of serial numbers on people in the Warsaw ghetto. He remembered the use of cattle cars to transport people to their deaths. He made the connection between the Nazis' treatment of Jews and the meat industry's treatment of animals. His experiences influenced him to found Farm Animal Rights Movement (FARM) and the U.S. movement for animal rights as a whole.[24]

Dr. Hershaft made me realize that animals are going through a holocaust every second of every minute of every day—and that I must use my voice and continue to raise as much awareness as I possibly can.

---

[24] *Holocaust To Compassion Alex Hershaft – Warsaw Survivor.* (2015.July.2015). Retrieved from https://www.youtube.com/watch?v=f7dZv43A0g0&app=desktop

As Elie Wiesel wrote, "We must take sides. Neutrality helps the oppressor, never the victim. Silence encourages the tormentor, never the tormented. Sometimes we must interfere."

WHAT SHOULD I DO?

The first thing to do is become as educated as possible. You can watch documentaries such as Earthings, Cowspiracy, The Cove, and Blackfish. Learn more about going vegan and becoming active by visiting www.KyleFerguson.net.

You can take action and get involved in local human and animal rights organizations such as Anonymous for the Voiceless. This organization has social media pages that you can visit to learn more.

EVERY INFLUENCE COUNTS

Admiral McRaven, the author of Make Your Bed, addressed the University of Texas at Austin Class of 2014 with an epic speech on the power of making a difference: What starts here changes the world. Tonight, there are more than 8,000 students graduating from U.T. Ask.com says that the average American will meet 10,000 people in their lifetime. . . . But if every one of you changed the lives of just 10 people, and each one of those people 10, and each one of those people changed the lives of another 10 people, and another 10, then in five generations—125 years—the class of 2014 will have changed the lives of 800 million people. . . . Think about it: over twice the population of the United States. Go one more generation and you can change the entire population of the world. Eight billion people. We have been given the miracle of life and we are the only species on planet earth who has a voice who can speak for the voiceless. Let's do our best to make the world the best place it can possibly be: a world of equality and caring for every being we share this planet with, the only world we know. Let's fill the world with love, gratitude, compassion, and happiness.[25]

---

[25] McRaven, W. H. (2014, May 19). University of Texas at Austin 2014 Commencement Address - Admiral William H. McRaven. Retrieved April 30, 2020, from https://www.youtube.com/watch?v=pxBQLFLei70&t=5s

ACKNOWLEDGEMENTS

I am grateful for the miracle of life, and I am incredibly grateful that I was born in Canada. Being born as a Canadian gave me the freedom and opportunity to travel the world, which opened my mind and gave me perspective.

Thank you to my parents, Scott and Karen, for allowing me to pursue my dreams in life.

Thank you to Anita Krajnc for your tremendous leadership. You helped to start a revolution with your act of compassion of feeding thirsty pigs' water. And you helped me find my life purpose to be a human and animal rights activist. And a voice for the vulnerable voiceless earthlings.

Thank you, Mark Cuban, for teaching me that time is the most precious asset, and you should be doing what is most important with time.

Your life lesson made me realize that it was an absolute must to complete this book.

Thank you, James Harbeck, for being a great editor. You did a great job of giving this book: structure, flow, and power.

Thank you, Derek Young, for being an amazing friend, and for giving me a kindle which led me to read and write relentlessly.

Thank you, Tim Keck, for being a great friend and providing wisdom and encouragement for me to write this book.

Thank you, Chris McGinn and Suzanne Gates, for being great friends. And thank you both for your indefatigable, utterly relentless, dedication to human and animal rights activism on the Light Movement.

Thank you, Peter Buffett, for helping me find my path to fulfillment and opening my mind with your wonderful book: *Life is What You Make It: Find Your Own Path to Fulfillment*.

Thank you, Jeff Bezos, for creating Amazon Kindle. And thank you for providing the opportunity to publish this book on your website.

Thank you, Nikola Tesla and Swami Vivekananda, for teaching me to do my best to make the world a better place each and every day.

Thank you, Charlie Tian, for giving me the chance to write for GuruFocus, which gave me the opportunity to practice writing.

Thank you, Nick and Tom Karadza, for teaching me the importance of building and maintaining momentum in life.

Thank you to the late and great Thomas Paine, for being a relentless human and animal rights activist.

Thank you, Leslie Dixon, for writing *Limitless,* and thank you to the entire cast of *Limitless* for doing a great job with the movie.

Limitless helped me expand my consciousness. *Limitless* is my favourite movie. And I watched it countless times as I completed this book.

I am forever grateful.

Thank you to all of the motivational speakers and people who I had the opportunity to listen to and keep me motivated as I faced adversity and hurdled obstacle after obstacle.

Tim Grover, Daymond John, Grant Cardone, Elena Cardone, Larry Ellison, Elon Musk, Gene Simmons, Ryan Holiday, Sara Blakely, Barbara Corcoran, Kevin O'Leary, Lauri Greiner, Robert Herjavec, Jerry Jones, Jennifer Lopez, The Grateful Dead, and Stefan Aarnio.

Thank you, Ben Patrick Johnson, for doing a great job of narrating this book, I am very grateful.

And thank you Kyle Maynard for being a great role model. And thank you for writing your awesome book: *'No Excuses'*. You taught me to be indefatigable, and that excuses are simply a way to avoid an obstacle without giving any effort to conquer it.

I am forever grateful for all of your wisdom and leadership.

Thank you as well to my parents, Scott and Karen, for giving me the miracle of life.

Cheers to filling the world with love, gratitude, peace, and happiness.

CAN YOU DO ME A FAVOR?

Thanks for reading Compassion Is Not a Crime: The Anita Krajnc Trial.

Would you please write a review about this book on Amazon?

I would greatly appreciate it!

Reviews are a great way to spread awareness and build momentum for the fastest growing social justice movement of our time.

Click here to leave a review on Amazon.com

Note: If the link does not work on your device, please visit Amazon manually and navigate to the Compassion Is Not a Crime: The Anita Krajnc Trial page, and leave a review.

Thank you – I really appreciate your support.